# Al-Dhabh

SLAYING ANIMALS FOR FOOD
THE ISLAMIC WAY

## Ghulam Mustafa Khan

Published by:
Islamic Medical Association
London and
Ta Ha Publishers Ltd
68a Delancey Street, London NW1

© 1982 Ta Ha Publishers

Joint publication with:
Ta Ha Publishers Ltd
68A Delancey Street, London NW1
and
Islamic Medical Association (UK and Eire)
London


**British Library Cataloguing in Publication Data**

Khan, Ghulam Mustafa
  Al-Zabah: slaying animals for food the Islamic
way.—2nd ed.
  1. Slaughtering and slaughter-houses
  2. Muslims—Dietary laws
  I. Title
  636.089′4    TS1960

ISBN 0 907461 14 X

# CONTENTS

Verily Allah has precribed proficiency in all
things. Thus, if you kill, kill well; and if you
perform *dhabh*, perform it well. Let each one of
you sharpen his blade and let him spare
suffering to the animal he slays.


Dedication

To Those Who Cherish the Prevention
Of Cruelty to Animals

To the Members of the Students Islamic Society,
Cranfield Institute of Technology, Bedford, England,
who pioneered the provision of *Halal* meat
in the Institute's Offical Canteen.

بسم الله الرحمن الرحيم

*In the name of Allah the Beneficient the Merciful*

## Introduction

*All praise is due to Allah, the Creator and Sustainer of all the worlds. All praise is due to Him Who has made man a steward on His earth and granted him guidance and sustenance.*

This is a revised version of a booklet that was first published in 1976 under the title "Al-Zabah" by Green Link Limited. It was based on a paper presented at a symposium of the Royal Society of Medicine on Humane Killing and Slaughter-house Techniques. The symposium was held at the Universities' Federation of Animal Welfare Associations (UFAWA) on 20 January 1971.

Since that time, the debate on humane methods of killing, far from subsiding, has been periodically stoked up, in the United Kingdom in particular, whenever Muslims apply to abbatoirs for licences to slay animals for food the Islamic way. The most long-standing and persistent attempt to alter the Islamic and indeed the identical Jewish method has come from the Royal Society for the Prevention of Cruelty to Animals, the RSPCA and its branches abroad. The RSPCA has been keen to portray *dhabh* as a painful, cruel and barbaric method of killing animals. It is one of the purposes of this book to show that such campaign to discredit the *dhabh* method is totally misguided and in fact without any sound and scientific basis. Moreover, the methods which are pressed upon the Muslims as clinical, modern and painless can be shown to be both painful to the animal and unhealthy to the meat consumer.

The promotion of healthy and correct patterns of food consumption is very much part of the Islamic scheme of things. It is the main purpose of this book to show that such patterns of food consumption

7

are achieved through precise Islamic dietary laws which specify what can and what cannot be eaten. It will be shown how these laws encourage a strong attachement to, and a respect for, the sanctity of life and an abhorrence to cruelty to mute animals which God has placed at our disposal for nourishment and other uses.

One of the major sources of pressure to have Muslims "modernize" their method of slaying animals for food does not spring from the love of animals and the desire to prevent or minimize pain as is often alleged. Rahter, it springs from predominantly economic considerations. The highly automated, capital-intensive, macro-production methods of the meat industry can have no patience or sympathy with the time-honoured, tradition-tested but slow and seemingly cumbersome method of *dhabh*, never mind how healthy and wholesome this method can be proved to be. Economic considerations, however, should never be made the paramount factor in determining what is best for man expecially where his food, his health or his religion is concerned.

This booklet is one of the first publication of the Islamic Medical Association of the United Kingdom and Eire which is concerned to revive the system of Islamic Medicine and Healthcare, a system which is based largely on simple and natural methods and which also emphasises the principles and techniques of preventive medicine. In particular, this booklet is envisaged as an exercise in discharging our duty as Muslims to prevent cruelty to animals inflicted for economic or other considerations, to protect the health of consumers of meat and to remind people of their responsibility to God for the kind and proper treatment of animals and their accountability for any cruelty inflicted.

I am grateful to the executive and members of the Islamic Medical Association for the help given in preparing this booklet; to members of the Federation of the Students Islamic Societies in the UK and Eire from whom a great deal was learnt while giving lectures on the subject; and also to many others, including my children, who helped me to clarify some of the statements in the booklet.

I also wish to thank the Islamic Foundation of Nigeria and the Zabadne brothers of the Light of Islam Trust in Kano for their kind assistance.

Finally, thanks are due to Mr. Abdul Wahid Hamid who edited and prepared the manuscript for publication.

May Allah guide us the straight way and grant us all that is good in this world and in the hereafter.

Ghulam Mustafa Khan

*January 1982/Rabi'al-Awwal 1402*
*London*

*Note on the pronunciation of the word 'dhabh'*

The 'dh' in the word represents the Arabic letter ذ . It is pronounced with the tip of the tongue sticking out and pressed against the upper teeth, similar to the position of the tongue when pronouncing the 'th' in the word 'this'. The last 'h' in *dhabh* represents the letter ح and is pronounced with a strong explusion of breath.

# Chapter One

## Lawful and Unlawful Meat According the the Shari'ah

All life, animal as well as human, belongs to Allah. Animals, however, have been created for the benefit of man, who has duties towards them, and is accountable to Allah for their proper treatment.

"He, Allah, has created man out of a mere drop of sperm; and lo! this same being shows himself endowed with the power to think and argue!
And He has created cattle for you: you derive warmth from them and various other uses; and from them you obtain food; and you find beauty in them when you drive them home in the evenings and when you take them out to pasture in the mornings. And they carry your loads to many a place which otherwise you would be unable to reach, without great hardship to yourselves. Verily, your Sustainer is most Compassionate, Merciful." (*The Qur'an,* 16: 4-7)

The fact that all groups of living beings owe their existence to Allah and therefore stand on the same footing is beautifully expressed in the verse:

"...There is no beast that walks on earth and no bird that flies on its two wings which is not Allah's creatures like yourselves. No single thing have We neglected in Our decree." (*The Qur'an,* 6: 38).
The above verses give some idea of the sanctity in which life is held in Islam. It is in keeping with this spirit that Islam, as part of its comprehensive guidance, stipulates how animals are to be treated, what animals can be used for food by humans and how their meat is to be made pure and wholesome.

11

A general directive on food, addressed to mankind as a whole, is contained in the following verse:

"O Mankind! Eat of what is lawful and good on earth and follow not Satan's footsteps for, verily, he is your open-foe and bids you only to do evil and to commit deeds of abomination..." (*The Qur'an*, 2: 168-9)

Here mankind is asked to partake of the vast stoe of good things which Allah has prepared for them. They are cautioned not to eat everything indiscriminately nor to deny themselves, in the manner of ascetics, of what is lawful and good.

Then comes a specific directive addressed to believers in particular:

"O you who have attained to faith! Eat of the good things which We have provided for you as sustenance, and render thanks unto Allah, if it is truly Him that you worship. He has forbidden to you only carrion, and blood, and the flesh of swine, and that over which any name other than Allah's has been invoked, but if one is driven by necessity—neither coveting it nor exceeding his immediate need—no sin shall be upon him for, behold, Allah is Forgiving, Merciful." (*The Qur'an* 2: 172-3)

Here Allah commands the believers to eat of the good things He has provided and encourages them to give thanks as is due. Then, four categories of food are explicitly forbidden:

1. Dead animals
2. Blood
3. The flesh of swine
4. Animals killed in the name of other than Allah.

These four categories of unlawful things are repeated in *surah Al-An'am,* verse 145, where the flesh of swine is described as "loathsome" and animals over which any name other than Allah's is invoked is described as a "sinful offering".

In yet another verse, the Qur'an goes further, specifying in greater detail what is forbidden.

"Forbidden to you is carrion, and blood, and the flesh of swine, and that over which any name other than Allah's has been invoked, and the animal that has been strangled, or beaten to death, or killed by a fall, or gored to death, or savaged by a beast of prey, save that which you yourselves may have performed *dhabh* on while it was still alive; and forbidden to you is all that has been slain on idolatrous altars." (*The Qur'an*, 5: 3)

In the above verse, there are ten categories of forbidden things. In fact, five categories that are added — animals that die from strangulation, from a violent blow, a headlong fall, being gored to death or savaged by a wild animal — all fall within the category of dead animals. Some of these categories are particularly relevant, as we shall see, to some of the modern methods of so-called humane killing which are employed in abbatoirs and which Muslims are being pressurized to adopt.

### Dead Animals

Animals that die of natural causes are almost universally regarded as unlawful and unfit for human consumption. Animals may also die of disease or from eating poisoned plants, and it is naturally unthinkable to consider their meat for food. Another reason why dead animals or carrion cannot be considered for food by a Muslim is that their death was not brought about for the purpose of providing food, and there was no opportunity to declare the intention, or *niyyah*, of taking the animal's life for the sake of food. Such intention is an essential part of preparing an animal for food.

Apart from animals dying of natural causes, the five categories listed above are declared to be unlawful.
*Animals that die of strangulation (al-munkhanaqah)*: Strangulation can be accidentally caused by the tightening of a rope around an animal's neck, or by the animal's head entering a constricted space. Strangulation can also be brought about deliberately by man through squeezing the neck of poultry for example, or depriving it of air by other means. It can also be brought about by modern methods of chemical gassing. All these methods if deliberately applied are cruel and unlawful according to Islam and the meat of animals killed in this way is unfit for human consumption.
*Animals that die from a violent blow (al-mawqudhah)*: Blows can be

13

inflicted by a stick or similar instruments until the animal is dead. In modern terms this could also include electric shocks. Killing through electrocution is unlawful. Blows could also be inflicted by the poleaxe or the pistol. The latter was, until recently, widely used for stunning animals before killing them. It is possible that a pistol shot could deal a fatal blow before an animal is slain.

*Animals that die from a fall (al-mutaraddiyyah)*: A fall could be from a high place resulting in death from a broken neck or concussion. A fall could also be into a well and death could be caused by drowning. Thus breaking an animal's neck or bringing on a fatal concussion through the use of the poleaxe or pistol would make the meat of such an animal unfit for consumption. Killing chickens in electrified baths may also be considered as bringing on death partly by drowning and is therefore unlawful.

The two other categories in this regard are animals that are killed by being gored to death (*an-natihah*) and animals that are savaged by a beast of prey. Provided there is some movement of a limb or other part of the animal's body, it is permissible to perform *dhabh* on them and their flesh then becomes lawful as food. *Dhabh* is recommended to put the animal out of pain and agony. Incidentally, it is permissible to use the skin, the bones, the hair and the wool of animals that are found dead, but their flesh is forbidden.

## Blood

The prohibition of consuming blood is especially important for our discussion on methods of killing and the advantages of the *dhabh* method.

One of the functions of the blood is to carry nutrient material to the tissue cells and bring back waste products of tissue metabolism. These waste products are harmful to the body, and are removed from the blood when it passes through the kidneys. Dissolved in water these waste products are thrown out of the body as a solution we call urine.

Blood carries organisms responsible for various diseases. These organisms circulate in the blood without the body manifesting any symptoms of the disease, a condition called sub-clinical infection. It is therefore harmful to consume blood. Also, if meat containing much blood in consumed, there is a potential danger of contracting diseases produced through the organisms in the blood. It is therefore essential that the method employed in killing an animal for food should ensure the maximum extraction of blood from the meat. This point will be discussed in detail in the next two chapters.

One of the greatest harmful effects of consuming blood and meat rich in blood is psychological. It may produce a carnivorous psychology, inducing wild and savage behaviour. Consuming blood is destructive of the pure, human nature.

### The Flesh of Swine

The is expressly prohibited and described as "loathsome" (*rijs*) in the Qur'an. The pig is also regarded as *khabithah* or unclean and unwholesome. "And He has declared unlawful for you all that is bad or unclean." (*The Qur'an*, 7:157). The flesh of the swine is also prohibited under Jewish law and early Christians likewise regarded it as unclean until the convert Paul gave a dispensation in its favour.

### Animals killed in the name of other than Allah

This prohibition emphasises again the religious and sacred nature of taking life for food. This prohibition is meant to protect the principle of *tawhid* or affirmation of the oneness and uniqueness of Allah. It is meant to purify beliefs and combat idolatry and all its manifestations at all times. It is Allah who has created man and subjected whatever is on earth to him. He gave man the permission to take the life of an animal for his benefit and for this he is required to mention the name of Allah at the time of performing *dhabh*. If the name of any other than Allah is mentioned, the permission to take life is void and the meat of such an animal killed deserves to be made unlawful.

In declaring foods lawful and unlawful, one of the main criterian of the Qur'an is whether foods are good and wholesome or not.

"He (the Prophet) makes lawful to them the good things of life and he forbids them the bad things." (*The Qur'an*, 7: 157)

Further details on specific animals that are lawful to eat and others that are forbidden are given in the *hadith* or sayings of the Prophet. Specific

15

animals are mentioned as lawful such as the camel, the cow, the buffalo, the goat, the sheep, the gazelle, poultry, horses, rabbits and so on. Specific animals that are mentioned as forbidden are donkeys and mules, all predatory animals and birds, such as the wolf, the lion, the dog, the leopard, the tiger, the cat, the eagle, the falcon and so on. Also forbidden are a number of creatures, whose destruction the Shari'ah has recommended such as the kite, the scorpion, the rat, etc.

### Hunted Animals

There are special regulations concerning animals that are hunted. Animals and birds can only be hunted for the purpose of providing food. It is not allowed to use animals for target practice or to hunt animals for trophies. The persons who are allowed to hunt animals for food are the same as those who are allowed to perform *dhabh* (see Chapter Two). Animals which are at hand and on which *dhabh* can be performed directly are not allowed to be killed with a hunting instrument like an arrow or by a hunting animal. If there is life in an animal that is shot by an arrow, spear or other instrument or overpowered by a hunting animal like a dog, then that animal must be slain by the *dhabh* method. The Prophet, peace be upon him, has said, "If you despatch your dog (to catch an animal), mention the name of Allah over it. If the animal is caught and you realize that it is still alive, then perform *dhabh* on it."

The hunting animal or bird (like the dog or the falcon) used must be properly trained to catch animals for their master and not for itself. The Prophet, peace be upon him, has said, "If you despatch the dog and he eats from the animal that is caught, then do no eat from the animal for he has caught it only for himself. If you despatch a dog and he kills an animal and does not eat of it, then you can eat, for the dog has caught it only for its owner."

The name of Allah must be mentioned when despatching a dog or shooting an arrow or wielding a sword. Allah has said in the Qur'an:

"They will ask you what is lawful for them. Say, lawful to you are all the good things of life. And as for those hunting animals which you train by imparting to them something of the knowledge that Allah has imparted to yourselves, eat of what they seize for you, but

16

mention Allah's name over it, and remain conscious of Allah. Verily Allah is swift in reckoning." (*The Qur'an,* 5:4)

The pronouncing of Allah's name is an essential condition for the lawfulness of meat killed by a hunting animal. Adiy, a companion of the Prophet, once said to him, "I despatched my dog (to catch an animal) and (eventually) I found another dog with it. I do not know which of them caught the animal." The Prophet, peace be upon him, replied, "Then you must not eat, for you only pronounced the name of Allah over your dog and did not pronounce the name of Allah over the other."

The hunting of all sea creatures is allowed. All sea creatures are lawful to eat providied that they live solely in the sea or are not poisonous or harmful in any way. They are lawful irrespective of whether a Muslim or a non-Muslim catches them and there is no need to perform *dhabh* on any sea creature. Creatures that live on both land and sea, like the frog, are forbidden.

We have seen that there are clear categories of food that are lawful and categories that are forbidden. As for foods over which there is no clear pronouncement in the Shari'ah and no text forbids it, such food is considered lawful on the basis of the agreed legal principle that what is nor forbidden is allowed. This principle is derived from the Qur'anic verse which says:

"And He (Allah) has created for you whatever is on earth, all of it." (*The Qur'an* 2:29)

It is also derived from the saying of the Prophet, peace be upon him: "Allah has made obligatory on you certain matters and do not therefore neglect them. He has imposed certain limits so do not transgress them. And He has remained silent over certain matters out of His grace towards you and not out of forgetfulness, so do not search or pry into them."

## Chapter Two

## The *Dhabh* Method
## And Its Advantages

*Dhabh* is an exact and clearly defined method for killing an animal for the sole purpose of making its meat fit for human consumption. The word *dhabh* in Arabic connotes purification or rendering something good or wholesome. The *dhabh* method is also called in Arabic *dhakaat* which has the original meaning of purification or rendering something complete.

Certain conditions must be fulfilled for *dhabh* to meet the requirements of the Shari'ah.

1. The person performing the act (the *Dhaabih* in Arabic) must be sane, whether that person is a male or female, a Muslim or "Kitabi". (A "Kitabi" refers to either a Jew or Christian. For a discussion on the lawfulness of the meat of animals killed by a Kitabi, see Appendix I). If a person lacks or loses his competence through drunkenness, madness or through being a minor who does not possess the faculty of discernment, anything he may slay is not *halal* or lawful. Not lawful also is the meat of an animal killed by an idolator or someone who has apostacised from Islam.

2. The instrument usd to perform *dhabh* must be extremely sharp to facilitate the quick cutting of the skin and the severing of the windpipe and the bloodvessels so as to enable the blood to flow immediately and quickly — in other words, to bring about an immediate and massive haemorrage.

The Prophet, peace be upon him, has said: "Verily Allah has

19

prescribed proficiency in all things. Thus, if you kill, kill well; and if you perform *dhabh*, perform it well. Let each one of you sharpen his blade and let him spare suffering to the animal he slays." The last phrase "and let him spare suffering to the animal he slays" is a translation of the Arabic *wa-l yurih dhabihatahu*. This can also be translated as "and let him comfort the animal he slays" This has been taken to imply treating the animal gently and even providing it with water and food before performing *dhabh*.

The Prophet, peace be upon him, is reported to have forbidden the use of an instrument known as the "Devil's Cord" (*Sharitah ash-Shaytan*) which killed the animal by cutting its skin but not severing the jugular vein.

3. The incision should be made in the neck,★ at some point just below the glottis and the root of the neck. (In the case of camels, the incision is made in the hollow of the neck at a point which is called in Arabic, the *labbah*. Camels are slain by a process known as *nahr* which means spearing the hollow of the neck which is the most practical method of slaying this animal).

The throat (*hulqum*) and the oesophagus (*mari'*) must be cut in addition to the jugular vein and the carotid artery (*wadajaan*). The spinal cord must not be cut. The head is therefore not to be severed completely.

★In cases of necessity where it is difficult or impossible to reach the neck of an animal, for example if its neck is trapped in a crevice, it is permitted to slay the animal by a wound in any part of the body. This is known as *dhabh idtiraari* or the dhabh of necessity.

4. *Tasmiyyah* or pronouncing the name of Allah by saying *Bismillah* — In the name of Allah — before cutting.
According to the jurist, Imam Malik, whatever is slaughtered and the name of God is not mentioned over it, it is *haram* or forbidden, whether one neglects to say Bismillah intentionally or unintentionally. According to the jurist, Abu Hanifah, if one neglects to say "Bismillah" intentionally, the meat is *haram*; if the omission is unintentional, the meat is *halal*.

According to Imam Shafi'i, whether one neglects to say "Bismillah" before a slaughtering intentionally or unintentionally the meat is *halal* so long as the person is competent to perform *dhabh*. This opinion is

20

based on a report by A'ishah, the wife of the Prophet, who said that some people came to the Prophet and said, "O messenger of Allah, some people have brought us meat and we do not know whether the name of Allah has been mentioned on it or not."The Prophet replied, "Mention the name of Allah on it and eat." A'ishah added that the people concerned were recent converts to Islam.

It is enough to state here that the above tradition in no way proves that the pronouncing of God's name is not obligatory in performing *dhabh*. In fact the tradition emphasizes that the pronouncing of God's name was a widely known matter and was considered an essential condition of *dhabh*.

*Abominable Acts in Slaying an Animal.*

1.   It is abominable to first throw the animal down on its side and sharpen the knife afterwards.
It is related that the Prophet once passed by a person who, having cast a goat to the ground, was pressing its head with his foot and sharpening his knife while the animal was watching. The Prophet said, "Will this goat not die before being slain? Do you wish to kill it twice? Do not kill one animal in the presence of another, or sharpen your knives before them."
2.   It is abominable to let the knife reach the spinal marrow or to cut off the head of the animal.
3.   It is abominable to break the neck of an animal or begin skinning it while it is convulsing or before its life is completely departed. The Prophet, peace be upon him, has said, "Do not deal hastily with the souls (of animals) before their life departs." (*Laa ta'jilu al-anfus qabla an tazhaq.*)
4.   It is abominable to perform *dhabh* with a dull instrument. The Prophet commanded that knives should be sharpened and that knives should be concealed from animals to be slain.

From the foregoing description, it can be seen that both faith and a precise method are conditions for the validity of *dhabh*. The insistence on pronouncing the name of God before slaying an animal is meant to emphasize the sanctity of life and the fact that all life belongs to Allah. Pronouncing the *tasmiyyah* induces feelings of tenderness and compassion and serves to prevent cruelty.

21

The actual method of *dhabh* has many advantages. To begin with, the speed of the incision made with the recommended sharp knife is a relatively painless process and initially in itself is a form of stunning. No additional stunner, mechanical or otherwise is necessary. The question of pain and *dhabh* is discussed at greater length below.

One of the main advantages of the method of *dhabh* is that it allows for the most rapid and efficient bleeding of the animal. It is known that since blood clots after death, it can only be removed when the animal is not dead. It is also obvious that blood being enclosed in a closed circuit can only be removed by cutting the blood vessels. The greater the number and larger the circumference of the blood vessels cut, the greater will be the amount of blood lost during the interval between the time the cut is made and when the animal finally dies.

The combined circumference of the jugulars, the two major blood vessels, and other small vessels make the neck the ideal place to cut and bleed the animal. The pressure in the arteries is the systaltic blood pressure receiving the direct thump due to the constriction of the heart. The force of the beating of the heart throws the blood into circulation. Therefore, the stronger the heart beat and the longer it goes on beating, the greater will be the quantity poured in to circulation, but the heart can eject only as much as it receives from the tissues. The rate and depth of respiration influences this. Another advantage of rapid and deep respiration is that it ensures adequate oxygenation and prevents stagnation or the increase in the acid base ratio of the tissue. This improves the keeping quality and the natural taste of the meat.

In order to squeeze all the blood out of the meat, the nervous connection between the brain and the body should be preserved and this is why the spinal cord is not cut in *dhabh*. Convulsions can only occur if this nervous connection is maintained. Convulsions occur in response to messages from the brain cells. Convulsions produce the squeezing or wringing action of the muscles of the body on the blood vessels which helps to get rid of the maximum amount of blood from the meat tissue into the circulation. The difference between the meat of the animal which does not convulse and that which undergoes convulsions, is like that between a wet sponge and one which has been

22

wrung out.

The physiological principles described above have a bearing on the removal of blood from an animal's body but they operate fully only if the animal is bled, while alive, by cutting across its throat and sparing the vertebral column without stunning the brain of the animal in any way.

We have in Chapter One above dealt with the prohibition on the consumption of blood according to Islamic law and we have seen how the technique of *dhabh* allows physiologically for the maximum amount of blood to be extracted from the meat of the animal. It is pertinent to state here that meat without blood tastes better and preserves better. On the other hand, the presence of blood in meat influences its putrefaction. Micro-organism find the blood a fertile ground in which to grow. The greater the amount of blood remaining in meat granules, the quicker will organisms proliferate and the sooner will the meat start putrefying.

## *Dhabh* and Pain

Having considered bleeding in *dhabh* and reached the conclusion that it is the most efficient method of bleeding, let us consider pain in *dhabh* in order to sort out the myths from the facts and determine whether it is *dhabh* that is painful or whether it is stunning or other techniques that are painful.

As the operation of *dhabh* is performed on animals that are mute, we have to use our own perception of pain in different circumstances to determine what pain is felt by the animal undergoing *dhabh*. In assessing possible pain, we have to consider the following steps in the procedure:

1. Cutting the skin of the neck;
2. The wound in the neck;
3. Bleeding;
4. Convulsions.

*Pain on cutting the skin of the neck*

Everyone who shaves his beard cuts himself at one time or another.

23

It is common knowledge that we do not perceive the pain from the cut till the wound starts bleeding or smarting. We feel this pain after the cut because the bleeding from the tiny capillaries is not severe enough to render us unconcious. The animals likewise would not feel the pain on the cutting of the skin of the neck by a very sharp instrument.

## Pain from the wound in the neck

Does the animal feel pain from the wound resulting from the cut in the neck as we do when we cut ourselves? The answer is no. In the operation of *dhabh*, the sharp knife, while sliding down the tissues in the neck, cuts open the four big blood vessels in the region. As a result, so much blood is lost so quickly that the animal becomes unconscious. Unconsciousness deepens as bleeding proceeds and the animal therefore cannot feel pain.

If the wound was inflicted on some other part of the body which could cause massive bleeding, the animal would remain conscious. Such a wound would be painful.

Any pain that the operation of *dhabh* might cause can be guessed by considering the similar operation of making a hole in the wind pipe — tracheotomy — performed in cases of respiratory obstruction. This operation can be done under local anaesthesia. In a planned operation (that is, when it is not urgent) the patient is given an injection of drugs to allay his sense of apprehension and of pain. This is supplemented by injecting a solution of local anaesthetic under the skin overlying the wind pipe, which is incised horizontally. In the animal, it is essential to cut the big blood vessels running vertically in the neck.

Though the surgical procedure in the two operations is similar, animal and man are not treated similarly not only because of the difference in pain sensitivity but also because of the different purpose of the operation. In the case of the animal, it is desired to bleed and kill and the whole operation is finished within a couple of minutes. For such a quick surgical procedure, the formalities that are observed in the case of a human being whose life is to be saved, are done away with. The animal does not need a tranquilizing injection and does not need a local anaesthetic. It is neither practical nor desirable to drug the meat of the animal. The knife serves the purpose of the injection and is more appropriate to be used on thick skin. As skin that is stretched cuts more easily, it is advisable to stretch the neck of the animal as is done with a human patient.

Man and animals have different sensitivity to pain. So far we have assumed animals to be as sensitive as humans, but this is not the case because the two structures concerned with pain, the skin and the brain, are not alike in man and animals. The brain of domesticated animals, like sheep and the cow, is anatomically and functionally different from the brain of man. Structurally, animals lack the frontal lobes which is well developed in man. The animal brain, in its natural state, is functioning like the brain of a human who has undergone frontal leukotomy or extirpation of the frontal lobes. Whereas a man with a sound brain is invariably conscious of the inevitability and significance of death, an animal lacks such apprehension unless it is badly handled and made to feel menaced.

The level of sensitivity to pain given to animals by their Creator is consistent with their natural behaviour. Blessedly, they have been created with a thick skin. They are created with a very low threshold and a very slight reduction of it will abolish pain. Therefore, going back to the discussion on pain from cutting the neck and the pain from the wound in the neck associated with a massive haemorrhage, it can be confidently said that an animal in this situation will not feel pain.

*Pain during bleeding*

Blood flowing from the wound of the neck of the animal slain by the method of *dhabh* gives an impression of cruelty to those who are ignorant of the physiology of pain. The process of bleeding is pain free and this can be confirmed by any blood donor. Furthermore, the animal's unconsciousness deepens as the bleeding proceeds.

*Pain and Convulsion*

Convulsions occuring in the animal lying with an open neck wound and with blood around, also impart an impression of pain and suffering. The method used therefore seems cruel to the eyes and mind of the onlooker who does not know the physiology of convulsions.

Convulsions are due to the contraction of the muscles in response to the lack of oxygen in the brain cells. The muscles, by these contractions, squeeze out blood from the blood vessels in the tissues to pout it into the central circulation system to be sent to the brain, but this is lost on the way and the brain cells consequently keep on sending messages to the muscles to wring out blood, until the animal dies. Convulsions thus occur when the animal becomes unconscious. Their occurrence confirms that the animal is unconscious. Those who have seen epileptics undergoing convulsion, falling and hurting themselves, will bear witness that the patients do not feel pain from the injuries they suffer during convulsions. They in fact do not even remember when they got hurt.

The convulsions, which are beneficial for healthier and tastier meat, are a loss to the meat trade. In an abbatoir the operator has to wait for the convulsions to die down before he can dress the animal. To save this waste of time from the point of view of the trade, immediately after bleeding, the spinal cord of the animal is destroyed. Consequently, the animal does not convulse.

*Role of the Vertebral Artery*

There has been some confusion on the role of this artery. The artery is so named because it runs through the tunnel formed by successive vertebrae in the neck. On each side of the neck, it runs from the root of the neck to above the first cervical vertebra, where the two arteries join to form the *baselar* artery in man. In cattle, the vertebral arteries do not supply the brain but end in the muscles of the head.

Even is this is not commonly known, and one assumes that the vertebral artery in man and animals supplies identical areas of the brain, it is widely known that a human can be rendered unconscious by compressing the carotid arteries from behind. The vertebral artery is not compressed and carries the normal load of blood, thus proving that it has no bearing on the onset of unconsciousness. Further, if mere compression of the carotid artery can produce unconsciousness, then section of both the carotids will induce unconsciousness that much faster. As already pointed out, we are not concerned with unconsciousness but with insensitivity to pain, and in man the vertebral artery does not nourish any of the areas which control common sensation.

## The casting of animals for slaying

Casting is the process of bringing animals down from a vertical (standing) position to a horizontal one (one its side) on the ground. The process is daily used by veterinary surgeons for the examination and treatment of animals, and nobody says a word. When the animals are cast for slaying, however, the process is labelled as cruel. This of course is applying double standards. There is nothing cruel or painful in the casting of animals for slaying.

## Pain and the technique of Dhabh

Preparation before the operation and the operative technique both influence pain. All good surgeons (operators) know this. Even a good technique will produce bad results in the hands of a bad operator. In such a situation, bad results are not the fault of the technique. The good operator knows and abides by all the rules and regulations, executes the operation humanely and in the spirit of doing his best.

Here, it is relevant to cite a rare observation noted by Tharton in his textbook on "Meat Inspection". It is alleged that animals have been seen to stagger after *dhabh* is performed. This surely will happen is someone tries to cut the neck of the animal with a blunt knife, or does not know how to use a sharp knife. This is an example of a bad operator but Thornton confuses it with the principles and technique of *dhabh*. If the knife is sharp and properly used, then, short of a miracle, the proper cut will be made.

Describing the factors which minimize pain during and after an operation "Wylle's Textbook of Anaesthesia" says rest, absence of hunger and thirst minimize pain, a sharp knife lessens pain. All this is in accordance with the teaching of the Prophet Muhammad when he spoke of performing *dhabh* proficiently, comforting the animal and sparing it suffering. The Universities Federation of Animal Welfare Associations, in their publication "Humane Killing of Animals" gives the following advice on the handling of animals:

27

"A quiet, unhurried approach and gentleness (is recommended). One should also try to make friends with the animal before killing it. Talking to it in a sympathetic way often helps, for the operation is far easier if the animal is not frightened, excited or apprehensive."

In the performance of *dhabh* all acts which would aggravate pain or interfere with the extraction of blood or both, are declared abominable. Such acts have been listed above and include cutting from the back of the neck, breaking the neck of the animal, cutting the spinal cord or using a blunt instrument. The principles and technique of *dhabh* thus show it to be an efficient, humane and pain free method.

# Chapter Three

## Alternative Methods

One of the main features of methods of slaughtering animals in Europe is the continual experimentation that has taken place. The result is that there has been no standard practice. Methods have changed rapidly in the recent past through the application of new devices and inventions and under the impact of the demands of the mechanized meat trade.

Prior to the twentieth century, various methods used were characterised by their extreme crudity and cruelty. Different methods were practiced on different animals. Some of these methods are described in detail in the book *Pistol versus Poleaxe* by Mcnaughten which was published in 1932.

*Sheep*: The sheep was laid on a couch and three legs were tied. The neck was extended and a sticking knife was then thrust through the neck behind the ears. With a second motion, the point of the knife was inserted between the joints of the vertebrae with the object of severing the spinal cord.

*Pigs*: There was no standard method used. Some slaughterers adopted the method of extending the pig's neck and making a vertical incision in the mid-line from the throat downwards. The pig has no identifiable neck making it difficult to incise horizontally. It screams frightfully on being held and continues to do so after the vertical incision is made as this does not induce massive bleeding. Therefore the pig remains conscious for some time after the incision is made. This combination of screaming with a bleeding throat cut has given the impression of cruelty. This has also been wrongly associated with cutting the neck horizontally with a sharp knife in other animals.

*Calves*:   Calves were hung upside down by their feet and the neck cut to obtain the bloodless meat called veal. This means that the significance of bloodless meat was known and appreciated.

*Oxen*:   Compared with oxen, other animals are small and, with the exception of pigs, easily controllable. Oxen could only be contraolled with great difficulty and required more time and more manpower. They were stunned by being hit on the head with a poleaxe or hammer.

Another method used on oxen was known as the English patent method which as originated in 1938 by a Dr. Carson. It consisted of fixing the animal either in a standing or recumbent position. Thereafter, the chest wall was punctured between the fourth and the fifth ribs and through the opening made, air was forced by a pair of bellows. The animal was suffocated by the pressure of air on the lungs, no bleeding operation followed. Though the method originated in England, it was better known on the continent. Other methods involving strangulation or the penetration of the brain with a heated spear were also adopted.

The methods used were time consuming and uneconomical from the point of view of commercialised slaughtering. New methods had to be found that were more economical but these turned out to be cruel also.

Around 1900, the era of mechanization was dawning in the west and the meat packing industry was one of the first to be mechanized. The realization that the methods used were cruel was growing especially since the formation in 1885 of the Model Abbatoir Society. The desire to change these methods gained impetus. There also emerged a third factor for change. In 1904, the Admiralty appointed a committee under Lord Lee of Farnham, which recommended that all animals be stunned before slaughter. This became a legal requirement in 1919. The non-mechanical stunner, the hammer or the poleaxe, was replaced by a mechanical one and after 1911, all animals big or small, had to be stunned by being shot in the head with a pistol.

### The captive bolt pistol

This was the first mechanical stunner introduced. For the purpose of slaughtering animals, the common pistol was made safer by captivating the bolt and so it became known as the captive bolt pistol. As a sales technique, the captive bolt pistol was named the "Humane Killer" to appeal to the emotions of laymen who generally assumed that shooting was much more humane than the older methods, especially the cutting of the neck which left blood around and the animal convulsing. The whole picture conveyed to them the impression of cruelty and they thought convulsions were a sign of pain. These false ideas were strengthened by contrasting it with the alleged advantages of the pistol.

Whatever the advantages of the pistol over the poleaxe, they were reaped by the slaughterer and his employer, the commercialized slaughter industry. From the point of view of pain, it was of no benefit to the animal. The pistol concusses and compresses the brain on the same principle as did the poleaxe. The pistol was eventually discarded by the very people who recommended it as humane in the first place.

## Electric Stunning

In 1933, the Slaughter of Animals Act was passed which required the licensing of slaughterers and the stunning of all animals with the exception of sheep and pigs if there was not a suitable electric supply. It was stipulated that the voltage used should not be less than 75 volts and that the current should be applied for not less than seven seconds. When stunning long-wooled breeds of sheep, the wool was required to be clipped from the sides of the head before the electrodes are applied, but few establishments have been prepared to spend time in carrying out this operation. Establishments are keen that as many animals as possible be stunned and killed in the least possible time and so commercial interests are often the overriding factor.

Is electric stunning painless? Scientists and physiologists have expressed serious doubts. Indeed, some are of the opinion that an electric current merely paralyses the animal and prevents it from uttering a sound but that the animal remains fully conscious and experiences great pain as the current is passed through the system.

31

They further object that if no special measures are taken for the current to be kept constant (as is the practice in the United Kingdom), and as the voltage is on the low side, a large proportion of animals with a high resistance to electronarcosis are not made unconscious at all, but are only subjected to partial paralysis, which deceives onlookers into assuming unconsciousness, where in reality there is consciousness which may be coupled with great pain. Those practicing electro-convulsive therapy (E.C.T.) on man are only too well aware of the unfortunate consequences of the accidental administration of an inadequate electrical dosage, which can permanently prejudice patients against further treatment.

In stunning animals the factors which make the inadequate administration of electric dosage a possibility are:

1. Variation in voltage;
2. Variation in duration;
3. Variation in resistance of the individual animals;
4. The human factor.

High voltage results in fractures of the bones. This depreciates the value of the carcass due to consequent haemorrhages. Low voltage results in motor paralysis and pain. Under slaughter-house conditions, fundamentally different from carefully controlled conditions, unintentional slips in the operation of electric apparatus are bound to occur.

A number of serious objections have been raised against electric stunning. It has been found that the vagus nerve is affected by stunning in a way that causes slowing of the heart and consequently there is less efficient drainage of the blood. Also, the haemorrages in meat due to shock convulsions make it impossible to differentiate them from haemorrhages due to diseases. In 1953, the Meat Inspection Branch of the United States Department of Agriculture came to the following conclusion:

"The use of electric stunning methods in plants which operate under Federal meat inspection has not been permitted as a result of experiments which were conducted several years ago at the University of Chicago. These experiments indicated that electric stunning in hogs

32

resulted in certain changes in the tissues which could not be differentiated by gross examination from similar changes produces by disease."

In 1955, the Danish Ministry of Justice issued a circular exempting the use of electric stunning for pigs. This was in response to a petition from Danish meat packers which said that

"Stunning with electricity causes extra vasation in meat, sanguinary intestines and fracture in the spinal column, pelvis, and the shoulder blades through shock. The blood in the meat makes it more susceptible to putrefaction and has a detrimental effect upon its taste. The properties of the meat which would cooperate with the salt in extracting the blood traces, are interfered with in the animal undergoing shock convulsions prior to slaughter."

In 1958, British regulations were amended and pigs were stunned by $CO_2$ gas, the reason being that stunning seriously affected the quality of British bacon and made it much less acceptable in competition with non-British varieties.

Electric stunning hastens the onset of putrefaction in meat. The explanation of the phenomenon lies in the high lactic acid level following electric shocks and prior to bleeding. High lactic acid alters the bacterial resistance of meat. (For a more detailed explanation, see E.H. Callow, *Food Hygiene*, Cambridge, 1952, p.42)

Experiences with electric stunning again show that seemingly efficient and trouble-free scientific methods are detrimental to both the animals concerned and the consumers of meat.

## Electrified Water Bath for Poulty Stunning

The Slaughter of Poultry Act was passed in May 1967 but was not brought into effect until 1970. The Act provides for the electric stunning of birds by an approved instrument prior to slaughter. The approved instrument, originating in Holland, is an automatic electric stunner based on the water bath principle. At the 1971 Royal Society of Medicine Symposium, a representative of the Universities Federation of Animal Welfare Associations (UFAWA), said that the Poultry Act required the bird to be rendered instantaneously insensitive to pain but

33

he himself felt unsure whether this was correctly interpreted under field conditions, because even inspecting officers have difficulty in differentiating electric paralysis from electric narcosis.

The problem with high voltage electric stunning for birds, where shackled birds are dragged over a high voltage grid, is that if the animal receives an inadequate dosage, it merely leaves the bird paralysed. High voltage, however, could kill the bird through bringing on heart failure and prevent effective bleeding of the carcass. The above problem is no nearer a solution even with the latest approved instrument. Here shackled birds are dragged over a bath containing electrified water. This method is, in fact, a combination of drowning and electrocution. UFAWA has come out against this process on the grounds that "drowning is slow and causes fear. Animals should never be drowned".

## Carbon Dioxide Stunning

After the disadvantages of electric stunning became known, it was given up in the USA (1953), in Holland (1955) and in Britain (1958). It was replaced by $CO_2$ (carbon dioxide) gas stunning. It is interesting to note that at the beginning of the century experiments with $CO_2$ were made on human beings but abandoned because it was not found to be humane. The meat trade was understandably interested in employing cheap gas on a large scale. In 1950, when reports of the disadvantages of electric stunning were probably becoming known, the use of gas was investigated by Hormel, a meat-packing firm in the USA. A successful technique was developed and the method came into use in Europe. The method is not used to any great extent with regard to sheep and calves.

The method involves a measured exposure of 65 to 70% gas and air mixture for a period of not less than 45 seconds, and requires bleeding which must begin within 30 seconds. In slaughter house conditions, the last two requirements are difficult if not impossible to practise. Even if strictly practised, $CO_2$ stunning is "chemical strangulation".

# Chapter Four

## Summary and Conclusions

The undeniable fact which emerges from the history of mechanical, electrical and chemical stunners is that after more than half a century of experimentation, there is not a single one that is safe to use, or than can be considered "humane" to animals, or that produces meat from which the maximum amount of blood is extracted.

So far as safety is concerned, it is always possible that in field xxxxxx

So far as safety is concerned, it is always possible that "in field conditions" the level of stunning that is applied either through mechanical or chemical means could be so great as to bring about the death of the animal before bleeding takes place. There is thus a danger of eating the meat of an animal that is dead before slaughtering.

One of the main arguments in favour of certain forms of stunning is that stunning reduces pain and is more humane to the animals concerned. In fact, as has been shown, stunning can be extremely painful. In recommending the "humane killing of animals", the Society for the Prevention of Cruelty to Animals has really accepted certain forms and degrees of cruelty to animals killed for food. Although electric stunning for example, has been shown to cause fracture in the spinal column, pelvis and shoulder blades of animals, the Society has been insisting that Muslims and Jews adopt this method for animals they slay. In recommending the water bath electric stunner for poultry, the Society is concerned with an apparent overall reduction of suffering. The conclusion is inescapable that stunning is favoured by people who find that paralysis of an animal which prevents it from screaming and convulsing, assuages their feelings when they are onlookers.

Concerning stunning and its effects on bleeding, it is a fact that all methods of stunning produce neurogenic shock, a condition in which blood leaves the circulation. In this condition, the nerves which regulate the size of the blood vessels are paralyzed. Blood fluid then leaves the circulation and enters the inter-cellular spaces in the tissues. When such an animal is bled, this fluid is not available for expulsion into the circulation and finally out through the wound. Moreover, since the brain is rendered inactive, convulsions in the animal's body is reduced to a minimum and blood remains in the meat tissue. As has been shown, the presence of blood in the tissue hastens the onset of putrefaction in meat and also has a detrimental effect on taste.

The only beneficiaries of mechanical, electrical or chemical stunning of animals is the meat industry. These methods make for a high "through put" in proceing animals. Using these methods, many more animals can be slaughtered in a given time and the economic advantages are therefore that much greater. It has been in the interests of commercial establishments to portray methods which involve stunning as humane. What is regarded as humane at one time is considered cruel at another time when other methods ensuring a more efficient and high "through put" are invented and introduced. When this happens, the big business lobby often manages to get legislation passed to protect their interests and also manages to secure the cooperation of pressure groups like the Society for the Prevention of Cruelty to Animals to support and champion the new methods. The advocate for mute animals thus wittingly or unwittingly becomes "a witness for the prosecution" against the interests of animals. It is interesting to note that the proponents of stunning accept the necessity of bleeding the animal through cutting the neck, but they propose stunning before bleeding on the assumption that stunning is painless and direct cutting with the knife is painful. This assumption of course is fallacious. When comparing pain caused by a knock on the head, electrocution or suffocation with pain from a cut by a sharp instrument (for example while shaving) we find that the latter is not even perceived when inflicted.

The proponents of stunning also assume that the methods now is use are essentially new and modern and are in sharp contrast with ancient and allegedly more barbaric methods. This of course is another fallacy. Animals that were stunned and strangled and animals that died through beating or through falling headlong from a height or being drowned (the meat of all of which is prohibited) represent animals that are shot in the head, gassed or electrocuted in the post-mechanical era. The processes and the end result are essentially the same. There is no difference between mechanical and non-mechanical stunners from the point of view of cruelty and hygiene.

In contrast to methods involving mechanical, chemical or electrical stunning, the *dhabh* method can and has been proven to be the best and most efficient one for slaying animals for food. It is an all-in-one-method. It produces shock due to rapid blood loss, it kills pain and it bleeds the animal thoroughly.

*Dhabh* ensures that a massive and rapid haemorrhage is brought about which is necessary to render the animal quickly unconscious and hasten its death. In order to squeeze out all the blood out of the meat, the nervous connection between the brain and the body must be preserved. It is mistakenly argued that the "Vertebral artery" which is not cut during *dhabh*, would delay the onset of unconsciousness. It is not conscioussness, but sensitivity to pain, which is most important in cutting the neck of the animal. Surgeons operate on conscious patients by abolishing this sensitivity to pain. The vertebral artery plays no part in pain sensitivity.

To the onlooker, ignorant of physiology, bleeding and convulsions appear as suffering. A blood donor does not feel pain when he is bled and an epileptic does not feel pain during convulsions. In fact, *dhabh* as the painless method of killing an animal by bleeding is similar to the process of bleeding a blood donor. The difference is that instead of a needle, a sharp knife is used to bleed. The knife quickly cuts the major blood vessels in the neck causing a rapid and massive haemorrhage. Thus, while 340 ml of blood is drained out of a blood donor in approximately 10 minutes, all the blood of the animal is drained in less time. In bleeding a blood donor, a small needle is inserted into the vein of the arm and the blood trickles through the needle while the donor is

37

conscious and remains conscious, until the bleeding is stopped.

*Dhabh*, as we have seen, is a complete and well-defined method, clearly defining the permissible and the non-permissible acts during its execution. It is an insurance against cruelty. In performing *dhabh*, the slayer is conscious of being accountable for his treatment of the animal to his and the animal's Creator. From the point of view of health and hygiene, it produces the best, the safest and the tastiest meat.

Unfortunately from the point of view of the mechanized and commercialised meat trade, the *dhabh* method of killing animals has a low "through put" and is therefore not as profitable. Commericial considerations, however, cannot and should not be allowed to ride roughshod over this natural way of preparing animals for food.

There are, of course, up-to-date scientific know-how and techniques that can be applied to improving the mechanics of the method and the conditions under which animals are killed especially in abbatoirs. Facilities for the transportation of animals, adequate place for rest, feeding and watering animals before *dhabh*, the manufacturing of sharp knives and devices for maintaining their sharpness, the construction of moving platforms to take the animal to the place of killing, clasps to keep larger animals under control and so on are examples of ways in which new developments can improve the conditions under which *dhabh* is performed.

In view of the above considerations, the attempt to discredit *dhabh* and the identical method adopted in Jewish practice (called *shechita* in modern Hebrew or *zebech* in old Hebrew) is ill-conceived. The related attempt to encourage Muslims to adopt various mechanical, electrical and chemical stunning methods is also patently misguided. Orthodox Jewish authorities have always stood firm in resisting these attempts while on occasions, there have been some persons who were prepared to state that there was no Islamic objection to the eating of meat from animals stunned before slaughter. Such opinions have come in 1928 from an Imam of the Woking Mosque in England, then under Qadiani control; in 1969, from an ex-Senior Kadhi of Tanzania and at some unspecified time from the staff of a certain (virtually unknown) Cadiz School in Cairo which is described by the RSPCA as 'a noted Islamic authority'. These 'authorities' are invariably cited by the RSPCA in their campaign against Muslim 'Ritual Slaughter'. (See Appendix III). The opinion of such 'authorities', it may be true to say, springs from an ignorance of what *dhabh* involves and one of the psysiological consequences of stunning before bleeding.

Based on the incontrovertible scientific evidence, it should not be too difficult for the Society for the Prevention of Cruelty to Animals to give up its ill-conceived and misguided attempts to persuade Muslims to give up the *dhabh* method and adopt the cruel and harmful practice of stunning animals. The *dhabh* method has stood the test of time and of scientific enquiry and remains by far the best, the most efficient, the safest and the most natural way to slay animals and make their meat fit for human consumption.

# Appendices

# Appendix I

## The Meat of Animals Killed by the Ahl al-Kitab

There has been among Muslims intense discussion on whether the meat of animals killed by the *Ahl al-Kitab* or People of the Book (by which is meant Jews and Christians) is lawful for Muslims to eat.

The Qur'anic verse in this regard states:
"This day all good and wholesome things have been made lawful for you. The food of those who have been given the Scripture (the Jews and the Christians) is lawful for you and your food is lawful for them." (*The Qur'an*, 5:6)

A few scholars in the past have interpreted this verse as giving an absolute freedom to Muslims to eat the food of the Ahl al-Kitab. One jurist, 'Ata', for example, states: "Eat of the animal killed by a Christian even if he says "In the name of Jesus" because God has given permission to eat the meat of animals killed by them knowing full well what they say." (Quoted in *Fiqh as-Sunnah* by Sayyid Sabiq, Vol 3, p.264). According to this opinion the Qur'anic verse which says
"And do not eat of what the name of God has not been mentioned over, for indeed that is an abomination (*fisq*)"
does not apply to Jews and Christians.

On the other hand, there is a body of opinion which has not been prepared to state that all food of Jews and Christians is absolutely and unconditionally lawful for Muslims. These include *sahabah* or companions of the Prophet like Ali, A'ishah and Ibn Umar. They have said that if you hear a *Kitabi* pronouncing the name of other than Allah the Almighty, then do not eat. (Quoted in *Fiqh as-Sunnah*, Vol 3, p. 264).

41

This latter view is naturally the more logical and reasonable. It is based on the reasoning that all food of the Ahl al-Kitab is lawful so long as it does not fall within the prohibited categories of food that are mentioned in the Qur'an and which we have dealt with in Chapter Two. The four broad categories of prohibited meat are carrion or dead animals (which includes animals that die from strangulation, from a violent blow, a headlong fall, being gored to death or savaged by a wild animal), blood, the flesh of swine and animals killed in the name of other than Allah. If the food of the Ahl al-Kitab includes any of these categories, then it mut be considered as forbidden or *haram*.

Based on current practice, only the meat of animals killed by orthodox Jews can be considered as lawful to Muslims. The Jewish method of slaying animals — *shechita* — is identical with the method of *dhabh* and does not admit of any strangulation or violent blow or any form of stunning. There are only a few minor differences between the Jewish practice and *dhabh*. In the Jewish method, the slayer (*sbochat*) is required to make the necessary cut in the neck of the animal in "one go". In the Muslim method, if the person raises his hand before completing the *dhabh* and then returns immediately to complete the process, this is allowed. In Jewish practice, only a specially appointed person is allowed to carry out the *shechita* whereas in Islam any sane adult Muslim who is acquainted with the process (pronouncing the name of Allah and knowing the parts of the neck to be cut) and the acts which are considered abominable is allowed to perform *dhabh*. Jews are not allowed to consume the meat of animals that have been injured but Muslims can, provided *dhabh* is performed on them.

Jews are only allowed to consume fat that adheres to the bones. They are not allowed to consume the hind quarters of the animal which includes all meat after the twelfth rib. Muslims can consume the meat procured by the mode of *shechita* which is practiced by orthodox Jews but which is not followed by reformed and liberal Jews. Where meat slain by a Muslim is available it has to be preferred.

A different attitude has to be adopted towards the meat of animals killed by Christians, particularly Christians in the West. Christians who have taken to the Pauline doctrine (and this includes Christians in Europe as a whole and Russia) have abandoned the Mosaic law. The

Pauline doctrine gave them permission to eat foods that were previously prohibited like the flesh of the swine. It also freed them from the particular method of slaying animals for food and gave them the option of adopting any methods available or devising ways of their own. The present situation is that among Christians, whether they might be practising or merely nominal Christians, the slaughtering of animals for food is not considered a religious act. It has been said that one of the reasons for the dispensation in the Qur'an allowing Muslims to eat the food of the Ahl al-Kitab (and have other relations with them) is that they are the closest to the Muslim believers in that they recognise revelation and the principles of religion. At the time of the Prophet, both Christians and Jews slaughtered animals according to the same principles as the *dhabh* method.

At present, however, the situation is quite different. Christians consume the flesh of swine and blood in the form of black pudding. Also, the methods that are used to slaughter animals generally requires various forms of stunning and these as we have seen involves the risk of eating animals that are dead before slaughtering or meat in which much blood remains. On all these counts, the meat of animals slaughtered by Western Christians cannot be recommended.

### Commentary by Abu'l 'Ala Maududi on the Qur'anic Verse

*"This day are all good things (tayyibat) have been made lawful for you. The food of those who have been given the Scripture is lawful for you; and your food is lawful for them."*

The words of this verse clearly point out that the only food of the People of the Book which has been made lawful for us is that which falls under the head of the *tayyibat*. The verse does not, and cannot, mean that the foods which are termed foul by the Qur'an and sound traditions and which we may not, in our own home or in the home of some other Muslim, eat or offer a Muslim to eat, would become lawful when offered us in a Jewish or Christian home. If someone disregards this obvious and resonable interpretation, he can interpret the verse in one of the following four ways only.

43

1.   That this verse repeals all those verses which have occurred in connection with the lawfulness and unlawfulness of meat in the surahs *an-Nahl, al-An'am, al-Baqarah* and in *al-Ma'idah* itself; that this verse of the Qur'an renders unconditionally lawful not only the poleaxed animal but also carrion, swine flesh, blood and the animal immolated to other-than-God. But no rational (*aqlee*) or transmissive (*naqlee*) evidence can ever be produced in favour of this alleged cancellation. The absurdity of the claim is shown by the fact that the three conditions of lawful meat occur in the surah al-Ma'idah itself, in the same context, and just before the verse now under discussion. These three conditions are:

1. It should not be the meat of the animals which have been declared to be unclean in themselves by God and His Prophet.
2. The animal must have been slain in the manner prescribed by the Shari'ah.
3. God's name must have been taken over the slain animal.

What right-minded person would say that, of the three consecutive sentences in a passage, the last would nullify the first two?

2.   That this verse countermands only slaughtering and taking God's name and does not alter the unclean nature of swineflesh, carrion, blood and the animal sacrificed to other-than-God. But we doubt if there exists, besides this empty claim, any solid reason for drawing a distinction between the two types of orders and for maintaining the one type and cancelling the other....

3.   That this verse fixed the dividing line between the food of Muslims and the food of Jews and Christians; that in the case of Muslims' food, all the Qur'anic restrictions would continue to be effective, but in respect of the food of Jews and Christians, no restrictions would obtain, which means that, at a Jew's or a Christian's, we may unhesitatingly eat what is presented to us.

The strongest argument which could be adduced in favour of this interpretation is that God knew what kind of food the People of the Book eat, and that if, having that knowledge, He has permitted us to eat their food, it means that everything they eat — including swineflesh, carrion, and the animal sacrificed to other-than-God — is pure and

44

lawful for us. But the verse on which this reasoning is based itself knocks the bottom out of this argument. In unambiguous terms the verse lays down that the only foods of the People of the Book which Muslims may eat are those which are *tayyibat*. And the word *tayyibat* has not been left vague; the two preceding verses explain at length what the *tayyibat* are.

4. That, out of the foods of the People of the Book, swineflesh alone may not be eaten, all other foods being lawful; or that, we may not use swineflesh, carrion, blood, and the animal slaughtered in other-than-God's name, though we may eat of the animal which has been killed in some way other than slaughtering and over which God's name has not been pronounced. But this interpretation is as unsustainable as the second.

No rational or transmissive argument can be given to justify the distinction between the injunctions of the Qur'an, to explain why, in respect of the food of the People of the Book, injunctions of one type remain in force while those of the other are rendered inoperative. If the distinction and the exception are grounded in the Qur'an, verses must be cited in proof, and if in the Tradition, the particular traditions must be referred to. And if there is a rational argument for it, it must be put forward.

## Juristic opinions

We shall now see that opinions have been offered by the various juristic schools on eating of the animal slaughtered by the People of the Book.

The Hanafites and the Hanbalites maintain that, for a Muslim, the food of the People of the Book is subject to the same restrictions which have been placed by the Qur'an and the Sunnah on the food of Muslims. Neither in our homes nor in the homes of Jews and Christians may we eat of the animal which is killed in some manner other than slaughtering and over which Allah's name has not been taken.

The Shafi'ites say that, since taking God's name is not obligatory, neither upon Muslims nor upon the People of the Book, a Muslim may eat of the animal which the Jews or Christians slaughter without taking Allah's name over it, though he may not eat of the animal which they slaughter in the name of other-than-Allah. The weakness of this position has been exposed above and so there is no need to discuss it here.

The Malikites, while granting that taking God's name is one of the conditions for the cleanness of the slaughtered animal, hold that the condition is not meant for the People of the Book, the animal slaughtered by them being unlawful even if God's name has not been taken over it. The only argument presented in support of this view is that at the time of the Battle of Khyber, the Prophet ate the meat sent by a Jewess, without enquiring as to whether God's name had been taken over it. But this incident could exempt the People of the Book from taking God's name only if it were established that the Jews of those times used to slaughter animals without mentioning God's name over them and that the Prophet, when he ate that meat knew that. To say simply that the Prophet did not ask whether God's name had been taken over it would not relax the condition in the case of the People of the Book. It is quite likely that the Prophet ate that meat unhesitantly because he knew that the Jews of his times took Allah's name over the animals they slaughtered.

Ibn Abbas says that the verse "The food of those who have received the Scripture is lawful for you" has repealed the verse "Eat not of that over which Allah's name has not been mentioned," and that the People of the Book have been exempted from observing this injunction. But this is Ibn Abbas's personal view and not a *marfu'* tradition. Moreover, Ibn Abbas is alone in holding this view, there being no one who is in agreement with him. Still further, Ibn Abbas does not offer any convincing reason as to why the one verse should cancel the other — and cancel only one verse and not the rest of the restrictions on food.

'Atu, Auza'i, Mak'hul and Laith bin Sa'd hold that the verse "The food of those who have received the Scripture is lawful for you" has rendered lawful "that which has been immolated to other-than Allah." 'Ata says that Muslims may eat of the animal slaughtered in the name of other-than-Allah. Auza'i says that one may eat of the game hunted by a Christian even if one hears the Christian taking the name of Christ

46

over his dog as he sets it off. Mak'hul says that there is no harm in eating of the animals which the People of the Book slaughter for their churches and synagogues and religious ceremonies.

But the only argument given in support of this is that God knew full well that the People of the Book sacrified animals in the name of other-than-God and yet He permitted the eating of their food. The answer is that God knew full well that the Christians ate swineflesh and drank wine, so why not make the verse declare lawful wine and swineflesh as well?

In our opinion; the soundest view is that of the Hanfites and the Hanabalites. Any other view one may hold is on one's own responsibility. But as shown above, the reasons and arguments advanced in favour of the views are so flimsy that, on the strength of them, the unclean cannot be proven to be clean, nor can the obligatory be made not obligatory. I would not advise any God-fearing person to adopt any of those views and to start eating of the animals slaughtered in Europe and America.

In the end, two clarifications are in order. Firstly, in killing small animals like the hen, the pigeon, etc., slight carelessness often results in an abruptly chopped-off head. Some jurists say that there is no harm in eating of such an animal. On the basis of this opinion, certain scholars have given the verdict that where a machine severs the head at one stroke, the condition of slaughtering is fulfilled. But to make the jurists' opinions into a basic law (*nass*) and derive from it rules which would alter the basic laws themselves is not a correct approach. The Shariah's injunctions about taking God's name have been given above, as have been the texts of the Qur'an and the Sunnah on which those injunctions are based. Now if the jurists have granted a concession in the case of an inadvertent violation of those injunctions, how can one regard this as the basic law and abrogate, virtually, the Shariah's injunctions about slaughtering? The jurists have said, and rightly, that one need not try to find out whether God's name has been taken over each and every animal slaughtered by the People of the Book; however, if it is positively learnt that, over a particular animal, the taking of God's name has been deliberately avoided, that animal may not be eaten. On the basis of this, again, it has been suggested that no inquiries need be made about the meat commonly available in Europe and America and that the animals slaughtered by the People of the Book may be eaten with the same ease of mind with which the animal slaughtered by Muslim butchers is eaten. But this logic would be valid only when we knew that a certain section or population of the People of the Book believe, in principle and as a matter of faith, that God's name ought to be taken at the time of slaughtering an animal. As for the people who we know are not at all convinced that a distinction between the clean and the unclean exists, and who do not in principle agree that taking God's or other-than-God's name makes any difference to the animal's cleanness or uncleanness, how can one take with an easy mind the animals slaughtered by them?

— *Tarjumanul Qur'an,* April 1959.

## Appendix II

### Statements in support of the Jewish method of slaying animals

*a.   Statement by Lord Horder, G.C.V.O., M.D., F.R.C.P.*

In January 1950 I was asked by the Board of Deputies of British Jews to give my opinion on the character of the slaughtering of cattle for food after the Jewish fashion.

I made careful observations of the process called *Shechita.* I reported as follows:

"The animal to be killed is isolated from the rest, placed in a padded pen which is rotated so as to bring the neck of the beast into position for the Schochet's operation. This consists in a clean and instantaneous cutting of all the blood vessels of the neck, together with the wind-pipe and gullet — in fact all the soft structures up to the spine.

"The animal loses consciousness immediately. It is difficult to conceive a more painless and a more rapid mode of death. For a few seconds after the cut is made the animal makes no movement. Its body is then convulsed; the convulsive movements continue for about a minute and then cease.

"The interpretation of these facts is clear. The cut is made by a knife so sharp and so skilfully handled that a state of syncope, with its associated unconsciousness, follows instantaneously upon the severing of the blood vessels, the rapid loss of blood and the consequent great fall in blood pressure. The movements of the animals, which begin about ninety seconds after the cut and continue for about ninety seconds, are epileptiform in nature and are due to the bloodless state of the brain (cerebral ischaemia with complete anoxaemia). Sensation has been abolished at the moment of the initial syncope.

"Careful and critical scrutinizing of this method of slaughtering leaves me in no doubt whatever that it is fraught with less risk of pain to the animal than any other method at present practised."

I was asked to repeat my observations with a view to a new statement which should be identical with this opinion or modify it if necessary. I made the new observations and I have no modifications to make in my original statement.

*b.    Statement by Leonard Hill, M.D., F.R.S., Director, Department of Applied Physiology, National Institute for Medical Research*

The Duchess of Hamilton's attack on the Jewish method of slaughtering animals is one calculated to raise prejudice. At the same time it is based on inaccurate observation and wrong deduction. It is an attack no less accurate and prejudiced than those frequently made on medical men who carry out scientific experiments of animals by leading anit-vivisectionists one of whom accompanied her on the visit to the Islington slaughterhouse.

There is nothing, so far as I know, which absolves a duchess, any more than one less exalted in rank, from the impartial examination of scientific evidence before making an attack in a public journal on a matter in which, judging by her article, she can have gained, through professional training, no adequate knowledge. A slaughterhouse is a horrible place to sensitive nature, whatever method of slaughter be used, and it is easy to arouse prejudice by descriptions of the scene therein, and by erroneous deductions made from observations of the movements made by the slaughtered animals after loss of consciousness. To the ignorant any sign of movement, and in particular movements which appear "purposive" in character, is taken as conclusive evidence of feeling, and yet we know that the body of even a decapitated animal will make such movements.

Several years ago I made a special study of the cerebral circulation, and later inquired into the methods of slaughtering at a time when the Jewish method was called into question by an Admiralty Committee.

All the evidence shows that complete cessation of blood-flow in the brain immediately abolishes consciousness in man, whether this be brought about by sudden compression of the carotid arteries in the neck, cutting of these arteries, or pressure applied to the brain. The very name "carotid" betokens the sleep which the ancients knew could be produced by compression of these arteries in a goat. Boys who accidentally kill themselves by playing at hanging do so because the pressure of the rope on these arteries suddenly deprives them of consciousness, and then they die of asphyxia, the weight of the unconscious body compressing the windpipe. Similarly it is very dangerous to breathe deoxygenated air because the loss of consciousness from want of oxygen is sudden and no warning sign is given. The brain loses its highest function, viz. consciousness, instantly on deprivation of oxygen, while all the lower functions of the nervous system and other organs continue to act for some time.

Now the Jewish method of slaughter consists in the sudden cutting of the neck right back to the bone, including the carotid arteries and jugular vein, the highly trained official using a very sharp knife. At once the whole of the blood is spilt out of the brain, and consciousness is abolished. No death could be more merciful, taking into account the fact that the animal, unlike man, has no knowledge or fear of impending death...

It is a certain fact that big injuries give no sense of pain at the moment of their infliction. This, long well known to surgeons, was abundantly proved in the Great War; men fight on, not knowing of their hurts. So, too, those who have cut their throats and recovered say they felt no sense of pain in the act of cutting. A horse can be bled from the jugular vein, if a sharp knife is used to make the necessary incision, while it quietly eats untethered at its manger, it does not feel the cut. The convulsive movements made by an animal after the brain has been

50

suddenly deprived of blood are caused by the excitation of the lower nervous centres by the sudden deprivation of oxygen. Such movements may be induced in man by compression of one carotid artery. I have done this on myself and have felt, to my astonishment, my arm making up-and-down movements and striking the arm of the chair. Of the nervous impulsion to movement I was wholly unconscious. All I felt was the arm rhythmically hitting the chair, and the feeling of faintness induced by cutting off half the blood brought to the brain by the carotid arteries.

As to the immediate cessation of the blood-flow in the brain when the carotid arteries were cut, I proved that this was so in the case both of unanaesthetized calves and goats. The blood pressure in the peripheral end of the carotid artery, that is the end in direct connection with the arteries in the brain, fell almost to zero in a second or two after cutting both carotids, a proof that the blood-flow in the brain had ceased within that period. It has been asserted that the vertebral arteries might convey blood to the brain and maintain consciousness after the cutting of the carotids. This cannot be so, for in the ox and the sheep the vertebral arteries do not supply the brain, but the muscles of the head.

Before cutting the throat by the Jewish method the animal is cast. A tackle is adjusted to three of the feet as soon as the animal enters the slaughterhouse. By hauling on this tackle the animal is made to roll over on to a floor on which according to the latest plan a large, thick mattress is placed.

According to the evidence of Mr. Openshaw, consulting surgeon to the London Hospital, confirmed by post-mortem examination, no bruising is produced by this operation, which is one so surprising to the animal that it cannot have any cognizance of the intentions of the slaughterer. A new method of casting has lately been introduced by Mr. J.R. Hayhurst, chief veterinary inspector at the Metropolitan Cattle Market. The development of this will still further lessen any discomfort caused by the operation.

After the casting the head is at once pulled back so as to stretch the throat, and then the Jewish official, who is a highly trained and conscientious individual, cuts the throat with one momentary sweep, and all is over. Some years ago, I carefully observed the slaughtering of a large number of animals by the Jewish and non-Jewish methods, both at Deptford and at Birkenhead. The Jewish method appeared to me humane and most certain and rapid in execution, especially in the case of wild and restive animals. The method of shooting by a free bullet is admittedly too dangerous to use in slaughterhouses.

In the case of the captive-bolt, the head has to be fixed before the bolt is shot into the skull. The fixation is difficult, and only possible in restive animals by roping them down. This particularly applies to animals imported alive, which come off the ships in a state wilder and more restless than that of home-grown oxen.

By proper organization all sight of slaughtered animals might be removed before the next victim is brought in, but I would add that I very carefully watched the behaviour of animals when brought into the slaughterhouse, and reached the conclusion that they were wholly ignorant of the death of their companions, whose bodies lay just dead, or cut up within their sight and smell. They were terrified of moving men, particularly of men moving in the shadows, and of being hustled by men this way and that. If no moving men were in sight they stood peacefully, and, I feel sure, oblivious of what, to sensitive humans, was the horror of the surroundings.

The Duchess of Hamilton has not justification for transferring her feelings of horror and consciousness of death, gained by spoken and written knowledge, to the beast of the field.

# Appendix III

## Campaign Literature issued by the Royal Society for the Prevention of Cruelty to Animals
### (Address: Causeway, Horsham, Sussex, RH12 1HG)

*a.   On "Humane Slaughter"*

For many years the RSPCA has campaigned vigorously to ban the practice of Ritual Slaughter, which is now permitted to certain religious communities under "The Slaughter of Animals Act", but without success.

These campaigns have failed to achieve their objective, even when a Private Members Bill has been before Parliament, for two main reasons:-
  (a)   There has been insufficient qualified evidence relating to pain, stress and efficiency of bleeding.
  (b)   Governments, and those religious communities concerned, have implied that such action could be considered as racial discrimination.

THE LAW

In the United Kingdom, "The Slaughter of Animals Act", 1958, now incorporated in The Slaughterhouses Bill which came into force on the 1st April 1974, states that all animals slaughtered in a slaughterhouse or knackers-yard must be:-
  (a)   Instananeously slaughtered by means of a mechanically operated instrument, or
  (b)   Stunned by means of a mechanically operated instrument, or an instrument for stunning by electricity, provided they are *instantaneously* rendered insensible to pain until death supervenes.

In all cases instruments must be in a proper state of repair. The slaughterman must hold a licence granted by the Local Authority and he must also be physically capable of the use of such an instrument without the infliction of unnecessary pain or suffering.

Pigs, however, may be anaesthetised by the use of cardon dioxide:

"The Slaughter of Pigs (Anaesthesia Regulations) 1958".

However, the Bill also states that a Local Authority shall not deny any religious community reasonable facilities for obtaining, as food, animal flesh from animals slaughtered by the methods specially required by their religion. Thus these provisions do not apply to a slaughter by the Jewish method for the food of Jews, by a Jew duly licenced for that purpose by the Rabbinical Commission or by the Mohammedan method for the food of Mohammedans and by a Mohammedan. However much we deprecate this, this is the state of the Law at present.

We are of the opinion that there must be something very wrong in a system whereby gentile slaughtermen have been prosecuted and fined for cutting the throat of an animal without previous stunning, while under the *same* roof some religious communities are allowed by Law to do exactly the same thing.

It is significant that the Member States of the EEC have accepted a Directive that food animals should be rendered unconscious before being bled to death. At the present time national authorities may grant special derogations in relation to certain religious rites, i.e. Jewish and Moslem methods of slaughter, but the very fact that the Directive has been adopted means that the vast majority of people in Europe agree that pre-stunning is more humane or less inhumane than cutting the throat of a fully conscious animal.

We are pleased to tell you that it was one of our Headquarters Veterinary Officers who was appointed to advise the Social and Economic Committee of the EEC on the question of Human Slaughter, that resulted in this Directive.

In the United Kingdom, an increasing number (there is no evidence of this—ed) of the more enlightened Moslem communities have accepted the opinion of the Imam of the Shah Jehan at Woking, that pre-stunning is permitted. This is in keeping with the Sacred Text which instructs Moslems to spare animals unnecessary pain.

If you wish, we would be grateful if you would send us any newspaper cuttings relating to Ritual Slaughter, i.e. proposals to use premises for that purpose, or equally where permission has been withheld by a Local Authority for the provision of such facilities.

Please write to your Member of Parliament and inform him of your concern about this aspect of animal welfare, because changes in the Law can only be effected by Parliament.

If you do write, please let us know, because this information could strengthen any attempt to get the Law amended.

b.  On "Ritual Slaughter for the Moslem Communities"

This Society, in common with the majority of veterinary surgeons and animal welfare societies, consider that greater suffering is caused by the Mohammedan method of slaughter, than by methods of slaughter carried out in accordance with the requirements of the Slaughter of Animals Act, 1958, and the Slaughter of Animals (Prevention of Cruelty) Regulations of 1958. These regulations include obligatory pre-stunning or anaesthesia.

Moslem slaughter can be performed by any follower of the Prophet, whether trained or not, with consequent disregard of the overall proviso that slaughter by the Mohammedan method must not inflict unnecessary suffering. It is an anomaly that Gentile slaughtermen have been prosecuted and fined for cutting the throat of an animal without previous stunning, while under the same roof persons of other religions are allowed to do exactly the same thing by law.

The consumption of meat by Moslems is permissible provided the meat is well bled, the appropriate prayer is said at the time of slaughter, or the pronouncement of the name of God is given before the meat is eaten if it was not pronounced at the time of slaughter. The other proviso is that the knife has not previously been used on swine. There appears to be no valid reason for excluding pre-stunning in Moslem slaughter, in fact it is permissible to eat the flesh of animals killed in the chase by a rifle, or other weapon provided always that the hunter is able effectively to bleed the animals and say the requisite prayer.

In 1969 the ex-Senior Kadhi of Tanzania pronounced in public there was no Koranic objection to the eating of meat from animals stunned before slaughter. His opinion was sought because of the introduction of captive bolt stunning in an abbatoir in Dar-es-Salaam.

The present Senior Kadhi of Tanzania subsequently endorsed the statement of the ex-Senior Kadhi. A similar pronouncement has been made by the staff of the Cadiz School in Cairo, which is a noted Islamic authority, and by the Imam of the Woking Mosque in England.

An increasing number of the more advanced Moslem communities have accepted the opinion of the Imam of the Shah Jehan at Woking that pre-stunning is permitted. This is in keeping with the Sacred Text which instructs Moslems to spare animals unnecessary pain.

## Appendix IV

### The Position the Irish Society for the Prevention of Cruelty to Animals (I.S.P.C.A.) and the Muslim Response

a.  *Letter from the I.S.P.C.A. to the Dublin Islamic Society.*

I.S.P.C.A.
1 Grand Canal Quay,
Dublin 2.
*28th October,* 1977

Dear Sir,

This Society, in conjunction with the RSPCA in England and Wales, the Scottish SPCA and the Ulster SPCA have for some years been concerned about the absence of pre-stunning of food animals before slaughter when carried out in the Muslim rite.

The present Senior Kadhi of Tanzania has endorsed the public statement of his predecessor that there was no Koranic objection to the eating of meat from animals stunned before slaughter, and a similar pronouncement has been made by the staff of the Cadiz School in Cairo as well as by the Imam of the Shah Jehan at Woking, England. My Society would therefore request an interview with the present leader of the Islamic community in Ireland to put forward our point of view and to learn from your Community if there are basic objections to pre-stunning before slaughter.

I am sure that a friendly and constructive discussion in this matter would help greatly to clarify the situation. I have been given the name of Dr. Mohammed Adam, but do not know if he is still resident in Ireland.

Yours sincerely,
W.E.P. Protheroe-Beynon
Administrator.

b.  *Dublin Evening Press Report on I.S.P.C.A.'s Campaign*

Evening Press,
*30th December,* 1979.

. . ."But," he (Major Protheroe-Beynon) said, "we have made absolutely no progress with the representatives of the Jewish community. I have made several approaches to the Chief Rabbi, Dr. Isaac Cohen, but I have got nowhere. And we can do nothing about it as the law stands."

However, he said he hoped Mohammedans in Ireland would soon accept that animals should be pre-stunned before killing in the Middle Eastern method, which involves cutting the animal's throat.

"That practice has now been accepted by their fellow Moslems in Britain, as a result of efforts by the RSPCA, and we are very much following the RSPCA line on this."

Major Protheroe-Benyon said he felt the only delay in getting Arabs here to agree to pre-stunning was in receiving the formal agreement of their religious leader in Ireland.

"But that office seems to rotate a lot — they come and go quite a bit; in fact they bring over a skilled butcher from the Middle East about once a fortnight to carry out their ceremonial killings," he said.

*c.   Letter to Evening Press, Dublin, from Ramlee Ismail, Vice-President, Dublin Islamic Society*

## "Islamic Way of Killing Animals is More Humane"

With reference to the Evening Press report on the humane method of killing animals for food, on behalf of my society, I would like to give our viewpoint regarding the matter and also to clarify certain points quoted in the article.

We Muslims in Ireland had been referred to as "Mohammedans in Ireland". That was an insult to us. We hope, in the future, those who do not have any knowledge concerning Islam will not make statements before consulting those who have knowledge of our society.

It was quoted that we bring a skilled butcher once a fortnight from the Middle East to carry out the so-called "ceremonial killings". That was not true. In fact, we have our own butcher, and the address is 7 Dunville Avenue, Ranelagh, Dublin 6.

It was also stated that the "Irish Society for the Prevention of Cruelty to Animals" is trying to persuade us to stun our animals before slaying them according to Islam. I would like it to be publicly known that our society rejects the idea of "stunning" the animals. We have medical proof to show that our method is most humane and healthy. Everybody is welcome to our centre for further information.

The proponents of stunning assume that stunning is painless (whereas anyone who receives a knock on the head knows that it is not).

Stunning is not a new conception, it was practised on uncontrollable animals like oxen. At present, it has simply been mechanised. The Europeans in the past stunned big animals by hitting on the head with a hammer. Was that humane? The pistol was the first mechanised stunner to be introduced in place of the pole axe (old method). The only advantage of using a pistol is its speed of action, thereby making it possible to kill a greater number of animals. For the animals, the knock by pole axe is as painful as the impact caused by the pistol. Thus, calling the pistol a humane killer is therefore a scientific distortion.

In 1933, electric stunning was introduced. A disadvantage of electric stunning is that it causes extra vasation in meat, sanguinary intestine, and fracture in the spinal column, pelvis and shoulder blades through shock.

Carbon dioxide gas stunning involves application of carbon dioxide gas measured exposure of 65-70% gas and air mixture for a period not less than 45 seconds and requires bleeding that must begin within 30 seconds of stunning, conditions impossible to attain and ensure under commercial conditions.

Our method of slaying the animals, which we call Zabah, is opposed to slaughter. It is a well defined — a divinely ordained method with sanctions, rules and regulations. Besides religious reasons, we also has phsiological , psychological and anatomical evidence to show that our method is humane, as opposed to the common practice, in this country in particular, and in Europe in general.

**Vice-President,**                                      **Ramlee Ismail**
**Dublin Islamic Society,**
**7 Harrington St., Dublin 8.**

## Appendix V

## Traditions of Prophet Muhammad Concerning the Treatment of Animals.

The Prophet, peace be upon him, forbade the killing of animals except for food. An-Nasa'i and Ibn Habban narrated that the Prophet, peace be upon him, said:

"Whoever kills (even) a little bird unnecessarily, it will complain to God on the Day of Resurrection and say, 'My Lord, so and so killed me in vain and did not kill me for a useful purpose'."

Muslim related on the authority of Ibn Abbas that the Prophet, peace be upon him, said "Do not take anything in which there is life or a soul (*ruh*) as a target."

The Prophet, peace be upon him, once came upon a bird which some people were using for target practice and he said, "God will disgrace whoever has done this."

Tirmidhi related on the authority of Jabir ibn Abdullah that a donkey once passed the Propher, peace be upon him. It had been branded on the face and blood was pouring from both its nostrils. The Prophet, peace be upon him, said "The curse of God be on the man who did this." Then he forbade (us) to brand animals on the face or hit animals on the face.

Yahya ibn Murrah narrated the following incident. He said: "I was in attendance on the Prophet, peace be upon him, when a camel came running and knelt down before him. Tears were flowing from its eyes. The Prophet, peace be upon him, commanded me to go and find out its owner. I left in search of the owner and found out that it belonged to a certain Ansari. I brought him to the Prophet who asked him, 'What is the matter with your camel?' He replied, 'I do not know why it is crying. We utilized its services, we saddled it with water bags to water date trees and gardens and now it is not fit for the job. Last night we decided to slay it and divide its meat among us.' The Prophet, peace be upon him, thereupon said, 'Do not slay it, Either sell it or give it to me'. The Ansari replied, 'O Messenger of God, accept it free of charge.' The narrator says that the Prophet branded it with the seal of the Public

60

Treasury (*Bayt al-Mal*) and included it among the animals belonging to the State. (This is evidence that in the Islamic State not only destitute humans but also destitute animals were cared for.)

Abu Hurayrah related (as recorded in al-Bukhari and Muslim) a saying of the Prophet about a man who was very thirsty and accidently found a well. He went down the well and drank some water. When he came out, he found a dog panting and licking the mud because of thirst. He remembered his own thirst and felt pity on the dog. He again went down the well and fetched some water for the day. God appreciated his kindness and forgave him. Those who were listening asked the Prophet if there was a reward for behaving kindly towards animals. The Prophet, peace be upon him, replied, "Behaving kindly towards any living soul is a blessing."

Abdullah ibn Umar and Abu Hurayrah related (as recorded in al-Bukhari and Muslim) that the Prophet Muhammad, peace be upon him, said that a woman was punished for killing a cat cruelly. She had tied the cat up. She neither fed it nor allowed it to feed on insects of the earth. Thus she starved it to death. As a punishment, she suffered punishment (in the hereafter).

Sa'd ibn Amr related (as recorded in Abu Dawud) that the Prophet, peace be upon him, passed by a camel whose belly was sticking to his back whereupon he said, "Be mindful of your duty to God in respect of these mute animals. Ride them when they are in good condition and slay them and eat their meat when they are in good condition."

# Islamic Medical Association of the U.K. and Eire

## Aims and objects:—

1  To serve humanity by
 — helping to prevent disease
 — treating the sick
 — promoting health habits and health education
 exclusively for the pleasure of Allah.

2  To revive preventive medicine and ensure that diagnostic and therapeutic medicine conform to the Shari'ah.

3  To promote medical education and health care by
 — setting up charitable clinics
 — providing medical supplies and literature wherever possible.

4  To organise relief work for areas hit by disease and/or disaster.

5  To encourage medical research and publications and establish an Islamic medical reference library.

6  To cater for the special medical needs of the Muslim community.

7  To encourage professional and social contact among Muslim doctors and help them wherever possible.

8  To promote better understanding, appreciation and practice of Islam.

## Publications of the I.M.A. and Ta Ha Publishers Ltd

1  Al-Dhabh: slaying animals for food the Islamic        Price £1.25

2  The Growing Threat: Smoking and the Muslim
 World                                                    £0.35

3  Muslim Dress        (under print)                     £0.45

4  A Deliberation of Food & Medical Products
 with substances of unlawful nature (Halal or Haraam)    £0.40

5  Personal Hygiene in Islam                             £0.40